MATTHIAS BURKE

CONQUEST: MY FIRST JEEP JLR

Hacks I've learned with a 2-door jeep and where it's taken me.

Copyright © 2024 by MATTHIAS BURKE

All rights reserved. No part of this publication may be reproduced, stored or transmitted in any form or by any means, electronic, mechanical, photocopying, recording, scanning, or otherwise without written permission from the publisher. It is illegal to copy this book, post it to a website, or distribute it by any other means without permission.

MATTHIAS BURKE asserts the moral right to be identified as the author of this work.

MATTHIAS BURKE has no responsibility for the persistence or accuracy of URLs for external or third-party Internet Websites referred to in this publication and does not guarantee that any content on such Websites is, or will remain, accurate or appropriate.

Designations used by companies to distinguish their products are often claimed as trademarks. All brand names and product names used in this book and on its cover are trade names, service marks, trademarks and registered trademarks of their respective owners. The publishers and the book are not associated with any product or vendor mentioned in this book. None of the companies referenced within the book have endorsed the book.

First edition

*This book was professionally typeset on Reedsy.
Find out more at reedsy.com*

Contents

1	Introduction	1
2	How and why, I chose the JLR	4
3	Welcome to the Jeep Life	9
4	To Modify or Not to Modify, is that really a question?	19
5	Ready to Go Camping and Trail Riding?	24
6	Conclusion	30

1

Introduction

W elcome to the world of Jeep ownership, where every turn brings a new adventure. I'm Matthias Burke, and I'm here to guide you through my experiences of owning a 2018 Jeep Wrangler Rubicon 2-door. In this book, you'll find practical advice, useful tips, and firsthand experiences to help you make the most of your Jeep journey.

Why this book? Because I've been in your shoes, and I know how daunting it can be to navigate the world of Jeep ownership. That's why I wrote this book—to share what I've learned, inspire fellow Jeepers,

INTRODUCTION

and give you the confidence to tackle any challenge that comes your way.

Throughout these pages, we'll explore everything from choosing the right Jeep to modifying it to suit your needs. We'll dive into the Jeep community, uncovering the camaraderie and support that comes with being a Jeeper.

So, if you're ready to unleash your inner adventurer and embrace the Jeep life, then let's hit the road and see where the Jeep life takes us next. Are you ready?

2

How and why, I chose the JLR

Choosing my Jeep Wrangler started with a simple question: What would I use it for? Was it going to be my go-to for tackling tough trails, or would it mainly be cruising around town? This decision mattered because it guided everything else. By thinking about how I wanted to use my Jeep, I could narrow down my choices and find the model that fit my lifestyle best. Whether you're into off-roading or just need a stylish ride for everyday life, figuring out your Jeep's main purpose is key to making the right choice. I decided that I wanted the best of both worlds.

2 or 4 Doors?

When I was faced with the decision between a 2 or 4-door, it was essential to consider how I was going to primarily use my jeep. The 2-door model offers a nimble and compact design, making it ideal for navigating tight trails and urban environments with ease. Its smaller size also contributes to better off-road maneuverability and a classic, rugged aesthetic. However, it comes with some trade-offs, such as limited cargo space and seating capacity.

On the other hand, the 4-door provides increased passenger and cargo space, making it a practical choice for families, road trips, and daily commuting. With its additional rear seats and larger rear cargo area, the 4-door model offers enhanced versatility and comfort for passengers and gear. While it may sacrifice some agility compared to its 2-door counterpart, the added convenience and functionality make it a popular choice for those seeking a more versatile vehicle.

Ultimately, the decision between a 2-door and 4-door depends on your specific needs and lifestyle preferences. Consider factors such as the number of passengers you'll typically transport, the amount of cargo you'll need to carry, and the types of terrain you'll be navigating. By aligning your choice with your intended use of the Jeep, you can ensure that you select the model that best suits your needs and enhances your overall driving experience. I bet you can guess which one I chose.

Soft or Hard Top?

When considering whether to opt for a soft top or a hard top for my Jeep, I had to weigh several factors, including my intended use of the vehicle and the road noise associated with each option.

The soft top appealed to me for its versatility and open-air driving experience. It's perfect for enjoying sunny days and immersing myself in the outdoor environment while off-roading or anytime I desired. However, I had to consider the downside: soft tops typically produce more road noise than hard tops, especially at higher speeds. While this wasn't a deal-breaker for me during off-road adventures or leisurely drives, it could become a nuisance during long highway trips or daily commutes, if it bothered me, I would simply turn up the music. I was a huge fan of being able to open my top while stopped at a traffic light (I don't recommend that).

On the other hand, the hard top offered superior insulation from external elements and reduced road noise, making it more suitable for highway driving and colder climates. The added protection from rain, wind, and noise made the hard top an appealing choice if I was seeking a quieter and more comfortable driving experience. However, I had to balance these benefits with the fact that the hard top is heavier and more cumbersome to remove, limiting the open-air driving experience (which I really wanted) compared to the soft top. I didn't want to have to constantly ask for help when I wanted to take the top off. This thinking was before I knew of all the ways to remove the top by myself, however I would've still chosen the soft top.

Ultimately, I prioritized open-air driving and off-road adventures, the soft top was the clear choice despite the potential for increased road noise. By carefully considering my intended use of the Jeep and weighing the trade-offs between the two tops, I was able to make an informed decision that enhanced my driving experience and enjoyment of the vehicle. In my opinion, soft top all the way.

Trim Level?

Selecting the right trim level for my Jeep involved a thorough consideration of my priorities and desired features. Each trim level offers a distinct blend of amenities, capabilities, and performance enhancements, catering to different driving needs and preferences.

The Sport trim: Serves as the foundational option, providing essential features for off-road adventures while maintaining affordability. With its rugged build and straightforward design, the Sport trim offers a blank canvas for customization and modification, allowing owners to tailor their Jeep to their specific preferences and needs.

Sport S trim: Introduces convenience and comfort features, enhancing the overall driving experience. From power windows and locks to air conditioning, the Sport S trim strikes a balance between affordability and additional amenities, making it an attractive choice for those seeking a more comfortable daily driver without compromising on capability.

Sahara trim: For drivers craving a blend of off-road prowess and upscale comfort, the Sahara trim level delivers. Equipped with premium interior features, such as leather upholstery, touchscreen infotainment systems, and enhanced sound insulation, the Sahara trim offers a refined driving experience both on and off the road.

Rubicon trim: In my opinion, the Rubicon trim reigns supreme. Built to conquer the most challenging terrain, the Rubicon trim comes equipped with specialized features such as heavy-duty axles, rock rails, and locking differentials. Whether tackling steep inclines or navigating rocky trails, the Rubicon trim offers unrivaled capability straight from the factory, making it my top choice.

As I mapped out the modifications for my Jeep, I aimed to strike a balance between enhancing its performance and appearance while keeping costs manageable. First, I outlined the key upgrades I deemed essential, such as lift kits, larger tires, and off-road lighting, focusing on improvements that would maximize both functionality and aesthetics.

To ensure I didn't break the bank, I scoured the aftermarket parts market, diligently comparing prices and seeking out discounts, promotions, and sales. This proactive approach allowed me to secure high-quality components at more affordable rates, enabling me to stretch my budget further without compromising on the quality of the parts. I knew that I wanted to take my rig off roading as much as possible and in order to make sure that I was not going to get stuck, the Rubicon was the way to go.

Moreover, I embraced the do-it-yourself (DIY) ethos, tackling installation tasks whenever feasible. By investing time in learning installation techniques and procedures, I not only saved on labor costs but also gained a deeper understanding of my Jeep's mechanics. This hands-on approach not only kept expenses down but also fostered a sense of pride and accomplishment as I personalized my vehicle to my liking. "Built, for me!"

Lastly, I prioritized modifications based on their impact and cost-effectiveness, focusing on upgrades that offered the most significant improvements within my budget constraints. This allowed me to tailor my Jeep to my preferences while ensuring that I maximized value for every dollar spent.

In essence, my approach to modifying my Jeep centered on making informed decisions, seeking out cost-effective solutions, and prioritizing upgrades that offered the most bang for my buck. By being resourceful and strategic, I was able to achieve my vision for my Jeep without overspending, ensuring that it was truly built my way while staying within budget, mostly. JEEP=Just Empty Every Pocket.

3

Welcome to the Jeep Life

Should I join a Jeep Club?

Joining a Jeep club is like making a group of friends who all love Jeeps just like you do. It's not just about having fun together, though that's a big part of it. When you join a club, you get to do cool things with your Jeep and learn a ton about owning one.

Imagine going on adventures with your Jeep, exploring new trails, and meeting new people who share your passion. That's what being in a club is all about. Plus, you'll get to participate in events and activities that you might not have known about otherwise. My local jeep club put on their own recovery classes and radio classes and even had "wrench turning parties."

But it's not just about fun and games. Being part of a Jeep club means you'll have a support system of experienced Jeep owners who can help you with any questions or problems you might have. You'll learn from their experiences and get tips on how to take care of your Jeep and make it even better.

So, joining a Jeep club isn't just about hanging out with other Jeep lovers. It's about becoming part of a community that shares your love for adventure, exploration, and all things Jeep.

Finding your local Jeep club is often as simple as a quick online search or reaching out to fellow Jeep owners in your area. Websites like JeepForum.com, Facebook groups, and forums dedicated to off-roading are excellent resources for connecting with local clubs and enthusiasts.

To determine if a Jeep club is a good match, consider the following factors:

Activities and Events: Research the types of activities and events the club organizes. Whether it's trail rides, off-road excursions, charity events, or social gatherings, make sure they align with your interests and availability.

Membership Demographics: Take a look at the demographics of the club's membership. Are they primarily focused on a specific age group, experience level, or type of Jeep? Choose a club where you feel comfortable and can relate to fellow members.

Club Values and Culture: Pay attention to the club's values and culture. Do they prioritize safety, environmental care, and responsible off-roading practices? It's essential to join a club whose values align with your own.

Location and Accessibility: Consider the club's meeting locations and how accessible they are to you. Choose a club with meetings and events held in convenient locations that you can easily access.

Reputation and Reviews: Research the club's reputation and read reviews or testimonials from current or past members. Positive feedback from fellow enthusiasts can be a good indicator of a well-established and reputable club.

By carefully evaluating these factors, you can find a local Jeep club that fits your preferences, interests, and values, ensuring a rewarding and enjoyable experience as part of the Jeep community.

What is the Jeep Wave?

Have you ever been driving your jeep had someone wave at you and you didn't know them? The Jeep Wave is more than just a casual greeting; it's a way of expressing camaraderie with others who share your love for the brand. It's a reminder that no matter where you are or where you are headed, you're never alone on the journey when you're part of the Jeep community. This unspoken tradition has deep roots in Jeep culture, dating back to the brand's early days. It's a tradition that transcends age, gender, and background, uniting Jeep owners from all walks of life under a common bond.

Participating in the Jeep Wave is not only about acknowledging other drivers; it's about celebrating the shared passion for exploration and adventure that defines the Jeep lifestyle. When you encounter another Jeep on the road, simply raise your hand or your fingers from the steering wheel, it's a simple yet meaningful gesture that fosters a sense of community and connection among fellow enthusiasts.

Who Let the Ducks Out?

The tradition of being "ducked" might seem quirky, but it's actually a fun and lighthearted part of Jeep culture. People started doing it as a way to add a bit of personality to their Jeeps and make them stand out from the crowd.

One story says it started with a Jeep owner who found a rubber duck in a parking lot and decided to attach it to their Jeep for good luck. Others liked the idea and started doing the same thing, and before long, it became a widespread tradition among Jeep enthusiasts.

Some people believe that the rubber duck represents good luck or serves as a mascot for their Jeep adventures. Others simply see it as a fun and whimsical way to express their personality and make their Jeep more unique.

Whatever the reason, it's a reminder that Jeep ownership is not just about practicality and performance—it's also about having fun and embracing the quirky traditions that make being part of the Jeep community so special.

What is Death Wobble?

Death wobble is a concerning and dangerous occurrence that can occur while driving a Jeep. It's characterized by a rapid and violent shaking of the vehicle, often felt through the steering wheel, which can make it feel like you've lost control of the vehicle.

Recognizing death wobble is crucial for ensuring your safety on the road. One of the most common signs is an intense shaking that typically occurs at higher speeds or after hitting a bump or uneven terrain. This shaking can be so severe that it feels like the vehicle is about to shake itself apart.

Preventing death wobble begins with regular maintenance and inspection of your Jeep's suspension and steering components. Ensure that all parts are in good condition, properly aligned, and tightened to the manufacturer's specifications. Pay attention to any signs of wear or damage, such as worn-out bushings, loose bolts, or damaged tie rods, and address them promptly to prevent potential issues that could lead to death wobble.

If you find yourself experiencing death wobble while driving, it's essential to remain calm and take immediate action to regain control of the vehicle. Begin by reducing your speed gradually while avoiding sudden movements of the steering wheel, which can exacerbate the wobble. Apply gentle pressure to the brakes to help stabilize the Jeep, but avoid slamming on the brakes, as this can cause further instability.

Once you've safely brought the vehicle to a stop, conduct a thorough inspection of the suspension and steering components to identify any potential issues that may have contributed to the death wobble.

By understanding how to recognize, prevent, and address death wobble, you can help ensure a safer and more enjoyable driving experience in your Jeep, allowing you to navigate rough terrain and obstacles with confidence and peace of mind.

I Can Get Free Oil Changes?

As a Jeep owner, you might be pleasantly surprised to learn about the perk of free oil changes offered by authorized dealerships or service centers. These complimentary oil changes are often part of maintenance packages or loyalty programs provided by Jeep. They help you save money while ensuring your Jeep's engine remains in top condition. Regular oil changes are crucial for engine health and longevity. By taking advantage of these free services, you can stick to the manufacturer's recommended maintenance schedule without any extra cost. Additionally, these programs may include other maintenance services like tire rotations and fluid top-offs, providing comprehensive care for your Jeep at no additional expense. To make the most of this perk, check with your local dealership or service center about available maintenance programs and eligibility criteria. It's a practical way to keep your Jeep running smoothly without breaking the bank.Top of Form

When to Get into 4x4?

Engaging four-wheel drive (4x4) mode is vital for better traction and control, especially off-road. To do it, know your Jeep's transfer case settings: 2WD, 4WD High, and 4WD Low. Make sure your Jeep is stopped and in neutral before switching.Sometimes if you feel like you are not able to switch between 2- and 4-wheel drive, put it into whatever gear you are trying to get into and then switch from N to R and then back.Sometimes that helps the transmission make the change

For slippery or uneven terrain like snow or mud, use 4WD High. This gives power to all wheels for better grip. Save 4WD Low for tough conditions like steep hills or deep mud. It gives maximum power for crawling over obstacles.

Remember, only use 4x4 when needed, and switch back on stable ground to avoid damage. Mastering 4x4 basics helps you handle off-road trails confidently and enjoy your Jeep's capabilities.

What Are Electric Lockers (Front and Rear)?

Electric lockers for the front and rear are crucial components that enhance traction and maneuverability, especially when navigating challenging terrain. These lockers work by locking the differential, ensuring that both wheels on the same axle spin at the same speed, regardless of traction conditions.

Front and rear electric lockers provide several benefits for off-road driving:

Enhanced Traction: By locking the differential, electric lockers ensure that both wheels on the same axle receive equal power, maximizing traction in slippery or uneven terrain.

Improved Control: Electric lockers help maintain stability and control by preventing wheels from spinning independently, reducing the risk of getting stuck or losing traction.

Better Maneuverability: With both wheels on an axle spinning at the same speed, electric lockers enable tighter turns and better maneuverability, allowing drivers to navigate obstacles more effectively.

Increased Off-Road Capability: Electric lockers are essential components for maximizing traction and overcoming challenging terrain such as rocks, sand, mud, and steep inclines. Their ability to lock the differential and ensure equal power distribution to both wheels on an axle significantly enhances the Jeep's off-road capabilities, allowing drivers to tackle obstacles with confidence and ease.

What is a Sway Bar?

A sway bar, also known as an anti-roll bar or stabilizer bar, plays a crucial role in stabilizing the Jeep's suspension system. It's a metal bar that connects the left and right sides of the suspension, helping to reduce body roll during cornering and maintain stability while driving.

Understanding the sway bar's function is essential for off-road driving:

Stabilizing Suspension: The sway bar limits the amount of body roll that occurs when the Jeep is turning, keeping it level and stable on the road. This helps improve handling and control, especially during sharp turns or sudden maneuvers.

Enhancing Off-Road Performance: While sway bars are beneficial for on-road driving, they can limit the Jeep's articulation and flexibility off-road. Disconnecting or adjusting the sway bars allows the suspension to move more freely, increasing wheel travel and improving traction on uneven terrain.

Knowing how to adjust or disconnect (electric or manual) the sway bars for off-road driving is crucial for maximizing the Jeep's performance:

Adjustable Sway Bar Links: Some Jeep models come equipped with adjustable sway bar links, allowing drivers to customize the sway bar's stiffness to suit their driving preferences. Adjusting the sway bar links can help balance on-road stability with off-road flexibility.

Sway Bar Disconnects: For more aggressive off-road driving, disconnecting the sway bars entirely allows for maximum suspension articulation and wheel travel. This improves traction and stability on rough terrain, enabling the Jeep to tackle obstacles with ease.

In summary, understanding the role of the sway bar in stabilizing the

Jeep's suspension system is essential for both on-road and off-road driving. By knowing how to adjust or disconnect the sway bars, drivers can optimize the Jeep's performance to suit various driving conditions and terrain types, ensuring a safe and enjoyable off-road experience.

What is Pitch and Roll?

Understanding pitch and roll is essential for off-road driving, as they relate to the Jeep's stability and maneuverability on uneven terrain.

Pitch refers to the up-and-down motion of the vehicle's nose and tail. In off-road driving, pitch degrees measure the angle at which the Jeep's front or rear end is tilted upward or downward relative to the horizontal plane. This can occur when driving on steep inclines or declines.

Roll, on the other hand, refers to the side-to-side motion of the vehicle. Roll degrees measure the angle at which the Jeep tilts from side to side while navigating uneven terrain or making sharp turns. Excessive roll can lead to loss of stability and increased risk of rollover.

Understanding pitch and roll is crucial for maintaining vehicle stability and safety while off-roading. Techniques for managing pitch and roll include:

Maintaining a Proper Center of Gravity: Keeping the Jeep's center of gravity low by evenly distributing weight and avoiding overloading can help minimize the risk of excessive roll and maintain stability on uneven terrain.

Using Proper Driving Techniques: When driving on steep inclines or declines, approach them at a controlled speed and maintain steady throttle and brake control to prevent sudden shifts in pitch. Similarly, when navigating sharp turns or obstacles, use proper steering and throttle control to minimize roll and maintain stability.

Utilizing Vehicle Features: Some Jeep models are equipped with features such as electronic stability control (ESC) and traction control systems, which can help mitigate the effects of excessive pitch and roll by automatically applying brakes or adjusting engine power to individual wheels.

By understanding the concepts of pitch and roll degrees and employing proper driving techniques, jeepers can effectively manage vehicle stability and navigate challenging terrain with confidence and safety.

4

To Modify or Not to Modify, is that really a question?

Should I give my Jeep a Name?

Giving my Jeep a name holds more significance than one might think and surprisingly it took me longer than I thought it would. It's about infusing it with personality and character, transforming it from a mere vehicle into a trusted companion on my adventures.

Naming my Jeep allows me to form a deeper connection with it, ascribing qualities and traits that resonate with its identity. Whether it's a reflection of its rugged durability, its adventurous spirit, or simply its unique appearance, the name serves as a symbol of its individuality and purpose.

Through this process, I not only personalize my Jeep but also infuse it with a sense of identity that aligns with my own. It becomes more than just a mode of transportation—it becomes a part of my journey, embodying the experiences and memories we share together.

What Size Wheels and Tires I Was Going to Use?

Selecting the right wheels and tires for my Jeep involves considering key factors like terrain, performance, and aesthetics. For rough off-road trails, larger wheels and tires with deeper treads offer better traction. However, they can affect handling and fuel efficiency on highways. Smaller wheels and tires may be more fuel-efficient but compromise off-road performance. Aesthetics also matter, as they contribute to the Jeep's overall look. By balancing these factors, I can choose wheels and tires that suit my driving needs and personal style.

I had already decided on what size wheels and tires I was going to use because of a friend of mine that had the size and look I was going for. My first set of aftermarket tires were a set of Cooper STT Pro 37/13.5R17.

The first number equals the height of the tire. The second number equals the width of the tire. The R means that it's a radial tire and the last number indicates the size of the rim. So, my first set of tires were 37 inches tall by 13.5 inches wide mounted on a 17-inch rim. I was getting ready for the trails.

What About Steering Upgrades Because of the Bigger Tires?

When upgrading to larger tires, it's crucial to consider the impact on the steering system. Larger tires exert more force on the steering components, potentially leading to increased steering effort and decreased responsiveness. Upgrading the steering system is essential to accommodate the added stress and ensure optimal handling and safety.

One common upgrade is installing a stronger steering stabilizer or steering damper. This component helps reduce steering vibrations and wobbles, providing a smoother driving experience, especially at higher speeds or over rough terrain. I went with a Falcon 2.0 steering stabilizer right from the beginning just to make sure that I didn't over work my steering

Additionally, upgrading to heavy-duty steering components, such as tie rods, drag links, and track bars, can enhance steering precision and durability. These components are designed to withstand the increased load exerted by larger tires, minimizing the risk of premature wear and failure.

By investing in steering upgrades, Jeep owners can maintain optimal handling and safety, even with larger tires installed. It's an essential consideration for anyone looking to maximize the off-road capabilities of their vehicle while ensuring a comfortable and controlled driving experience in various conditions. I eventually added a 2.5-ton JL HD 2" Aluminum Steering kit from RPM Steering and the American Iron Offroad ball joint delete kit. Death wobble is a real thing and shouldn't be taken lightly.

What Additional Lighting Did I Want to Add?

I decided to make my Jeep safer and give it a cooler look by adding some extra lights. One type of light I chose was the KC Cyclone rock lights. They might be small, but they're really powerful. These lights help me see under my Jeep when it's dark outside, which is super useful, especially when I'm driving off-road on rough trails.

Another type of light I picked is the Gravity LED Pro6 lights. These lights are bright and can shine really far ahead. So, when I'm driving on trails at night, these lights make sure I can see everything clearly in front of me. They're like my extra pair of eyes, making me feel safer and more confident when I'm out exploring.

Apart from adding lights, I also made some changes to the front fenders of my Jeep. I decided to chop the fenders and install American Adventure Labs High-Line Fender Braces. These braces come with daytime running lights, which not only make my Jeep look awesome but also serve a practical purpose. By removing the inner fenders, gutting the stock fender liner and adding these braces, I created more space for the front wheels to move freely, giving my Jeep better clearance. It's like giving my Jeep a makeover while making it more capable on the trails.

Where Am I Going to Store Everything?

When planning for off-road adventures, camping or both, it's essential to consider storage solutions for all the gear, tools, and accessories you'll need and want. This includes both interior and exterior storage options to accommodate various items effectively.

Interior storage options might involve setting up overhead organizers, seat-back organizers with MOLLE webbing, or cargo nets to secure smaller items and make them easily reachable. Additionally, utilizing cargo liners or organizers for the cargo area can keep larger items organized and prevent them from shifting during off-road journeys. Personally, I found a large container sturdy enough to accommodate all my recovery gear, tool roll bag, snatch blocks, and an extra winch line for emergencies. To keep this container secure while traveling, I use a ratchet strap to hold it in place. Additionally, my seat backs have MOLLE webbing where I've attached my first aid kit and some extra storage pouches for small items, ensuring easy access to essential supplies while on the go.

When it comes to exterior storage, options such as roof racks, cargo carriers, or hitch-mounted cargo baskets provide extra room for storing bulky gear like camping equipment, coolers, or spare tires. These solutions help clear up space inside the Jeep while keeping necessary items within reach. Fortunately, I received a Garvin Expedition Rack as a Christmas gift from my amazing wife. One of the reasons I love this rack is its ease of detachment from the Jeep when it's not in use.

By exploring storage options, you can ensure that your gear is well-organized and readily accessible, enhancing the enjoyment and ease of your off-road adventures. Platforms like Amazon, Facebook Marketplace, and local Jeep club FB buy/sell pages can be valuable resources for finding the ideal storage solutions tailored to your specific needs.

5

Ready to Go Camping and Trail Riding?

Let me tell you about my very first camping trip with my Jeep and the first time I tried rock crawling with it. It was such an exciting adventure! But along the way, I learned some really important things that I want to share with you.

One of the biggest lessons I learned is about safety. Rock crawling can be risky, especially if you're alone and not fully prepared. So, I want to make sure you understand that it's not a good idea to go rock crawling by yourself unless you're really confident in your skills and you have all the right recovery gear to help you out if something goes wrong.

Another thing to consider is the skills of the people you're going with. It's important to make sure everyone knows what they're doing and can handle the challenges of rock crawling safely. It is implied that if you are going with a group, normally the person that is "hosting" the trip is very experienced. There are usually people there with varying experience from novice to rock crawler extraordinaire, and all the people that I have gone with are very helpful without being overbearing or judgmental.

When it comes to picking where to go rock crawling, choosing the right Off-Road Park can make a big difference. Look for parks with good facilities and amenities that can help you stay safe and have a great time.

Lastly, when you're getting ready for your trip, remember to pack smart. Only bring along what you really need to stay safe and have fun. By keeping these things in mind, you'll be well-prepared for your own Jeep camping and rock crawling adventures!

How Much Does It Cost?

When it comes to planning or joining a group going on a Jeep camping trip and trail riding adventure, it's crucial to think about the financial aspects involved to ensure a smooth and enjoyable experience. Here are some of the various expenses you'll encounter:

Firstly, consider the significant portion of your budget that will go towards fuel costs. Your Jeep will need gas to cover the distances to and from your destination, as well as any driving you'll be doing while exploring trails. Depending on the distance you plan to travel and the current price of gas, fuel expenses can add up quickly. Also consider what you have mounted on top of your jeep. The extra gear will cause a significant amount of drag and really reduce your MPG.

Next, food expenses are another important consideration. Whether you're planning to cook meals over a campfire, bring along pre-packaged food, or dine out at local eateries, budgeting for groceries or dining expenses is essential. Consider the duration of your trip, the number of people you'll be feeding, and any dietary preferences or restrictions when planning your food budget. I have always brought what I wanted to eat and then a little more just in case, but every time we've camped out with a group, we all ended up sharing food, grills, whatever we had so everyone had an enjoyable time.

Camping fees or Off-Road Park entrance fees are also something to account for, especially if you'll be staying at a designated campground. Many campgrounds charge a nightly fee for campsites, which can vary depending on factors such as location, amenities, and time of year. Research campground fees in advance and factor them into your budget to avoid any surprises upon arrival. Most if not all off the parks that I have visited have allowed camping on the park grounds, just think about the time of year and if you will need reservations.

Considering and budgeting for these various expenses, you can ensure that you have the financial resources necessary for a successful Jeep camping trip and trail riding adventure. Planning ahead and being prepared financially will allow you to focus on enjoying the experience without any financial stress or limitations.

What to Pack and Where to Put It on the Jeep?

When preparing for a camping and trail riding excursion with your Jeep, it's essential to pack wisely and efficiently. Here's a guide on essential items to bring along and practical tips on how to organize and secure gear on your Jeep for safe and efficient transport:

Essential Items: Start by packing essential camping gear such as a tent, sleeping bags, camping stove, cooking utensils, and food supplies. Don't forget to include personal items like clothing, toiletries, and any medications you may need. For trail riding, bring trash bags so you can make sure that you leave the trail better then you found it, pack recovery gear such as a tow strap, traction boards, a tire repair kit, and a snatch block or two. It's also wise to bring along tools for basic vehicle maintenance and repair.

Organization: Think about how you are going to eat and drive. If you have a soft sided cooler put some snacks and drinks in that and keep it close to you. This way you are not having to stop or try to get into your big cooler with all the food for your meals. Depending on how you have your gear packed, think about the accessibility of any emergency gear. Make sure that is the easiest to get to and not buried under other gear.

Secure Your Gear: Ensure that all gear is securely strapped down or stored to prevent shifting or falling during transit. I am a firm believer of keeping everything that is inside the jeep secure so that there are no flying objects in case of a quick stop or while you are out on a trail and you are going up a steep rock shelf or trail.

Weight Distribution: Distribute weight evenly throughout the Jeep to maintain balance and stability, especially when traversing rough terrain. Place heavier items low and towards the center of the vehicle to prevent tipping or overloading.

By carefully packing and organizing your gear and securing it properly on your Jeep, you can ensure a safe and enjoyable camping and trail riding experience without worrying about lost or damaged equipment.

What Difficulty Trail Am I Really Ready to Tackle?

When picking which off-road trails to try, it's important to think about a few key things for a good experience. First, be honest about your off-roading skills. If you're new, stick to easier trails. Next, think about what your Jeep can handle. If it's got modifications, it might handle tougher trails better. Also, consider what you're comfortable with. If you're not confident with steep or tricky terrain, go for simpler trails. Some times the group might break up into ability groups and ride the appropriate trails. There is no shame in starting slow, even if you are a beginner with a modified jeep. Get the confidence then go for the harder trails. You really don't want to out drive your skill level and end up rolling your jeep.

What Recovery Gear Do I Need and Do I Have All My Recovery Gear?

When venturing off-road, having the right recovery gear is essential for handling emergencies effectively. Here's what I think you need to know:

Gloves are a must-have to protect your hands during recovery tasks. Traction boards provide traction to get unstuck from slippery or muddy terrain. Tow straps are crucial for pulling vehicles out of tough spots, so make sure you have one with a high tensile strength and suitable length. Winches are powerful tools for self-recovery or assisting other vehicles, and it's vital to ensure yours is properly mounted and maintained. Soft shackles are safer alternatives to metal shackles, reducing the risk of injury or damage during vehicle recovery. Kinetic ropes, also known as snatch ropes, provide dynamic pulling force to free stuck vehicles safely. Additionally, snatch blocks can be used to increase the pulling capacity of your winch and change the direction of the pull, adding versatility to your recovery toolkit.

Before hitting the trails, ensure all your recovery gear is properly packed and secured in your Jeep. Regularly inspect and maintain your equipment to keep it in good condition and ready for use when needed.

With the right recovery gear, you can handle emergencies confidently and enjoy your off-road adventures to the fullest.

6

Conclusion

My continuing journey with Conquest, my 2018 JL Rubicon has been nothing short of exhilarating. From the moment I made the decision to purchase it to the unforgettable adventures it has taken me on, every experience has been filled with excitement, challenges, and valuable lessons. Throughout this book, I've shared my insights into selecting the right Jeep, making modifications, embracing the Jeep life, and embarking on unforgettable camping and trail riding expeditions.

I hope that my experiences and advice shared in this book have provided valuable guidance to fellow Jeepers, whether they are considering their first Jeep purchase or seeking inspiration for their next off-road adventure. Remember to prioritize safety, be prepared with the right gear, and always respect the trails and the environment.

Thank you for joining me on this incredible journey. Your support and enthusiasm mean the world to me. If you've found this book helpful, I would greatly appreciate it if you could take a moment to leave a review on Amazon. Your feedback will help other Jeep enthusiasts discover the

CONCLUSION

joys of Jeep ownership and off-road exploration. Until we meet again on the trails, keep on Jeepin' and embracing the spirit of adventure.

www.ingramcontent.com/pod-product-compliance
Lightning Source LLC
Chambersburg PA
CBHW070957220526
45471CB00007B/3064